YOUR KNOWLEDGE HAS VALUE

Bibliographic information published by the German National Library:

The German National Library lists this publication in the National Bibliography; detailed bibliographic data are available on the Internet at http://dnb.dnb.de .

Imprint:

Copyright © 2015 GRIN Verlag, Open Publishing GmbH
Print and binding: Books on Demand GmbH, Norderstedt Germany
ISBN: 978-3-668-11470-8

This book at GRIN:

http://www.grin.com/en/e-book/312467/biometric-voting-system-in-ghana-a-case-study

Chukwuka Anowu, Thompson Oyetunji

Biometric Voting System in Ghana. A Case Study

GRIN Publishing

GRIN - Your knowledge has value

Since its foundation in 1998, GRIN has specialized in publishing academic texts by students, college teachers and other academics as e-book and printed book. The website www.grin.com is an ideal platform for presenting term papers, final papers, scientific essays, dissertations and specialist books.

Visit us on the internet:

http://www.grin.com/

http://www.facebook.com/grincom

http://www.twitter.com/grin_com

Biometric Voting System In Ghana

Anowu D. Chukwuka

Department Of Computer Engineering

All Nations University College

Koforidua, Ghana

Oyetunji T. Thompson

Department Of Computer Engineering

All Nations University College

Koforidua, Ghana

Abstract—Presently in Ghana, the electoral process involves biometric registration of eligible voters and queuing up of the registered voters, to cast their votes using ballot papers and indelible ink on the day of voting. At the end of the voting exercise, the counting of the ballots begins by the electoral officials, manually which consumes a lot of time, in order to obtain the election results and is also vulnerable to human errors. This project allows the casting of electronic votes from various polling stations of the Electoral Commission of Ghana in a controlled environment. The system maintains the biometric data, which is the thumb print and finger print of eligible voters which will be used for verification on the day of voting. On the day of voting, the voting portal will then be activated by the verified voters by swiping their voter's ID card at the barcode scanner, so as to enable him/her cast their vote at the voting portal made available by the system. The system employs real-time collation of election results which reduces the time taken in collating election results. The project aims to give an insight into the efficient management of electioneering processes, a system which operates in a timely manner, with minimum bureaucracy.

Keywords— EMS - Election Management System; DRE - Direct Recording Electronic System; DES - Data Encryption Standard; EVM's - Electronic Voting Machines; PC – Portable Computer;

TABLE OF CONTENTS

I. INTRODUCTION 3

II. RELATED WORKS 4

III. PROPOSED WORK 5

IV. PERFORMANCE EVALUATION 7

    A. FINDINGS 7

    B. GRAPHS AND EVALUATIONS 8

    C. ADVANTAGES OF THE PROPOSED VOTING SYSTEM OVER THE EXISTING VOTING SYSTEM 10

V. CONCLUSIONS AND FUTURE SCOPE 11

VI. LIMITATIONS 12

REFERENCES 13

I. INTRODUCTION

"An Electronic voting system is a voting system in which the election data is recorded, stored and processed primarily as digital information" [1]. As democracy is being implemented in our present day government, voting becomes a very crucial activity, especially as voting is a fundamental human right. So in having a free and fair election, we want to ensure that each vote that is casted, is counted and recorded with accuracy and impartiality.

Many citizens of Ghana especially the eligible voters suffer from poor officiating and management of voter's information and elections, coupled with the lengthy period of time taken in the collation of election results.

We believe that this occurs as a result of the corrupt practices of government officials and prominent individuals in elections, and the vulnerability of human errors occurring during the electoral processes such as counting of the votes, which is done using manpower and subsequent time delay in the collation of results.

We want an efficient and effective management of electoral practices and voter's information, in which the collation of election results is automated and done in real-time.

The objectives of this project includes the following,

- To develop a software that will enhance effective electioneering processes.

- To reduce the occurrence of human errors in capturing the voter's data during the registration process.

- To develop a software that will drastically reduce the expenses incurred from carrying out elections, by eliminating the use of ballot papers.

- To reduce the occurrence of human errors in the collation of election results, by speeding up the counting process and calculation of election results.

II. RELATED WORKS

The research on e-voting is very essential for the progress of democracy within a nation. According to A. Al-Ameen and S. Talab "If a secure and convenient e-voting system is provided, it will be used more frequently to collect people's opinions through cyberspace" [2]. Other researchers have made big contributions and improvements in voting systems, especially in electronic voting systems, which includes F. A. Haziemeh *et al*, whose project is aimed at developing a secured electronic voting system which will prevent the casting of votes twice and also disallow people who are not eligible to vote from casting votes [3].

V. K. Yadav *et al* proposes a voting system, named KBEVS model which provides a comfortable facility for busy electorates to cast their votes. The electorates use their mobile phones to cast their votes [1].

M. A. Musa and F. M. Aliyu proposes a voting system that "will provide the voter/user the opportunity to cast his/her vote where ever he/she has access to the internet or intranet… The voter can as well monitor the real time situation of the registration process. The system has the capability to check the validity and eligibility of a voter, thereby blocking invalid votes and access of illegitimate users to the systems" [4].

According to A. Al-Ameen and S. Talab, "in some societies, like in the developing countries, not all voters have access to a computer and internet… However, if the election is only facilitated by internet voting, then the technology would end up becoming a barrier to voter's participation" [2].

The limitations in the e-voting system by M. A. Musa and F. M. Aliyu includes, "that it is an internet application and not a standalone software. The other limitations includes that the software cannot be installed on computers and moved around with. The software only works with MySQL database and the software only works on a server with a PHP engine" [4].

The e-voting system proposed by F. A. Haziemeh *et al*, is very vulnerable to errors such as assigning "incorrect ID number to the citizen of a country… The user or voter cannot be allowed to see the results of the voting, until after the voting ends" [3].

The problems and limitations associated with the Estonian Internet Voting System, by the researchers which performed analysis during the 2013 local elections in Estonia, revealed a large number of poor security practices, which includes "attackers being able to maliciously alter the Estonian voting software during the build process, if it's created on a personal computer" [5].

III. PROPOSED WORK

Fig. 1 shows the architecture of the proposed system to be developed, indicating the components that are being implemented in the voting system, from one activity to the other in a sequential manner. This system records votes with the use of a PC machine to display the ballots, which can be activated by the voter. The proposed system uses electronic ballots and transmits the vote data from the polling station to the constituency level, and finally to the national level through a network (preferably an intranet). The vote data are transmitted as individual ballots are being cast. However, because transmitting vote data over public networks relies on equipment's beyond the control of the Electoral Commission of Ghana, the system is subject to additional threats to system integrity and availability [6].

The components that are implemented in the voting systems are as follows,

VOTER'S REGISTRATION

Every eligible voter is required to register on a day assigned solely for the purpose of registration, in which the individual will provide relevant information such as the first name, middle name, last name, date of birth and gender. This obtained information will be saved to the EMS, and a voter's number is automatically generated and assigned to each individual.

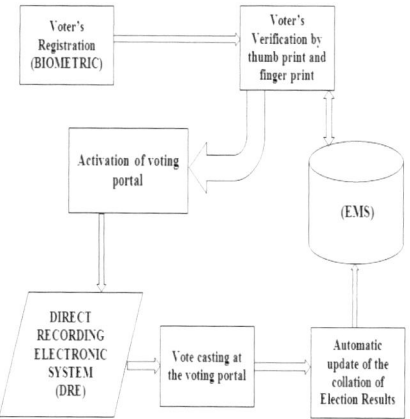

Fig 1 – Architecture of the Proposed System.

With this acquired information, a voter's ID card will be printed having all the voter's information and voter's number with a barcode embedding the voter's information.

After which, the voter's biometric data will be obtained, in which both the thumb-prints and finger-prints will be acquired and adequately saved alongside with the voter's information into the biometric verification system (EMS).

VOTER'S VERIFICATION BY THUMB-PRINTS AND FINGER-PRINTS

On the day of voting, the eligible voters which have been registered will undergo verification to authenticate they are indeed the individuals they claim to be, and to also ensure that the information in the EMS corresponds to the voter's ID card. This verification is done by identification of their thumb-prints and finger-prints, in which these biometric data exists in the EMS, if the voter's registration was successful.

DIRECT RECORDING ELECTRONIC SYSTEM (DRE)

The Direct Recording Electronic system is part of the Biometric Voting System, which records the voter's choices or selections made. The DRE is usually a touch screen device where the voter casts his/her vote [7]. The DRE is a PC machine, on which the biometric voting system is being implemented. The DRE's are networked both within the polling station and from the polling station to the constituency level and finally to the national level, for information flow or data transmission of the voter's information and election data.

ELECTION MANAGEMENT SYSTEM (EMS)

The EMS is the system responsible for the initialization of the components that collects the votes and also for the final tallying of the votes [7]. The EMS is usually located at every polling station, constituency, regional and national level and is implemented as a software running on the electoral machine (PC). The EMS incudes the web-based application and the desktop-based database.

REAL TIME COLLATION OF ELECTION RESULTS

"Josef Stalin once said: "Those who cast the votes decide nothing. Those who count the votes decide everything". The purpose of cryptographically verifiable voting systems is to prevent incorrect recording and tallying, by making the processes verifiable for everyone" [8]. This is why 3-DES is used as the cryptographic technique. As each voter casts his/her vote at the voting portal of the EMS, the election results at each polling station becomes automatically updated as the vote is recorded, also the constituency and the national level, also becomes automatically updated with the vote which has been casted, thereby collating the election results in a matter of seconds.

Other components of the proposed system that are being implemented in the system are fully covered below,

BIOMETRIC AUTHENTICATION

The biometric authentication is done with a thumbprint/fingerprint scanner, which will interface with a computer using a USB port. For the proposed system, we made use of the Suprema's award winning fingerprint algorithm.

VOTER'S ID CARD

The voter's ID card is a type of token-based authentication, where the token is the material containing both the voter's information and barcode, that the user possess. The voter's ID card is needed by the user and a card reader, which is the barcode scanner, which will be connected with a computer using a USB interface [8], so as to activate the voting portal on the Direct Recording Electronic system (DRE).

IV. PERFORMANCE EVALUATION

- DOMAIN – Elections.
- PLATFORM – Microsoft Visual Studio and Microsoft SQL server management studio.
- PROGRAMMING LANGUAGES – C# programming language, SQL and ASP.net.
- OS PLATFORM – (.NET Framework 4) Microsoft Windows XP and 7.

A. FINDINGS

During the project work, we decided to carry out a survey in order to find out the perceptions and opinions of 20 people in Ghana, when using the existing voting system, and the proposed voting system. We issued out questionnaires to eligible voters, who are citizens of the nation, of which 45% of the participants were males and 55% of the participants were females. The age group of the participants follows the following division which covers 65% of the citizens which belonged to the age 18 – 25 years, 10% of the citizens belonged to 26 – 33 years, 34 – 41 years and 42 – 49 years, and finally 5% of the citizens belonged to the 50 – 57 years. Of the citizens which filled the questionnaire, 80% had participated in the previous national elections. The questionnaire allowed us to generate data regarding parameters which are considered by the Citizens of a given voting system, which includes, their take on the existing voting system, Time consumption of both the existing and proposed voting system, simplicity and security of the proposed voting system, acceptance and future use of the proposed voting system.

B. GRAPHS AND EVALUATIONS

With the information obtained from this questionnaires, we performed our analysis, and the observations made, gave us the following results,

The existing voting system was thought to be time consuming by 90% of the participants and the remaining 10% found it not time consuming. The proposed voting system was viewed by 85% of the participants as simple, 80 % of the participants found the proposed voting system not to be time consuming. Some of the participants seem to think that the proposed voting system processes to be somewhat complex and unsure if the proposed voting system will be entirely secured. But, 80% of the participants also found the proposed voting system to be secured, meaning they can trust the proposed voting system and its capabilities.

Graph 1 - Analysis Of The Proposed Voting System.

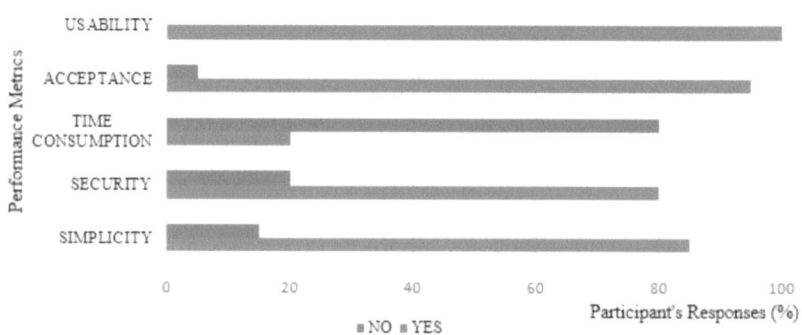

Fig. 2. A bar chart showing the respondents opinions of the proposed system.

8

Also, 95% of the citizens accepted the proposed voting system and when asked questions regarding the future use of the proposed voting system, 100% of the participants were willing to use the proposed votingsystem in the nation's next forthcoming election if it were to be implemented at such time. This shows the degree of satisfaction with using the proposed voting system. Fig. 2, below is a bar chart to demonstrate the analysis of the observations made from our findings.

Also from our findings, 95% of the participants preferred the proposed voting system over the existing voting system, after considering the previous aforementioned measuring parameters. This indicates there is a clear support for the proposed voting system, in this sense, this conclusion is equally optimistic, regarding the future use of the proposed voting system. The participants were excited about the automatic collation of the election results in the proposed voting system. The bar chart in Fig. 3, demonstrates the analysis of the observations made from the findings regarding the participant's preference of a voting system.

From the analysis of the observations made in our findings, 20% of the citizens finds the existing voting system to be good and 80% of the citizens finds the existing voting system to be bad.

The participants considered time consumption a major issue regarding the existing voting system. Fig. 4, below is a pie chart which demonstrates the participant's opinion of the existing system.

Finally, from the analysis of our findings done, the proposed voting system was clearly being supported by most of the respondents who participated in the survey. We can conclude that the proposed voting system is more beneficial in conducting elections, and also be optimistic that it will be accepted by the citizens of Ghana.

Graph 2 - The Respondents Voting System Preference

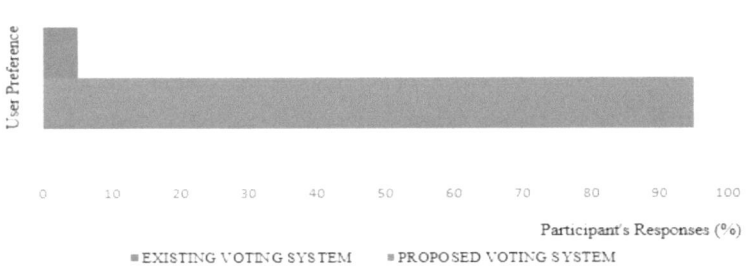

Fig. 3. A bar chart showing the respondents voting system preference.

9

Graph 3 - Respondent's View Of The Existing Voting System

Fig. 4. A pie chart showing the respondent's view of the existing system.

C. ADVANTAGES OF THE PROPOSED VOTING SYSTEM OVER THE EXISTING VOTING SYSTEM

1) Security: This existing system is vulnerable to fraudulent practices, such as ballot box tampering and manipulation of figures for number of votes. As for the proposed voting system, there are no ballot boxes, and the algorithm used to ensure the secrecy and counting of votes is difficult to break.

2) Double Vote: Double voting regularly occurs in the existing system. Due to poor checks applied. The occurrence of double votes in the proposed voting system is impossible due to the authentication and verification measures implemented.

3) Impersonation: Impersonation of electorates occurs in the existing voting system, whereas, in the proposed voting system, impersonation of an electorate is impossible, due to biometric checks and verification.

4) Voter's Turn Out: Due to the use of electronic technology, voter's turn out will be increased during elections.

5) Human errors: Human errors occurs in voter's registration process and collation of election results for the existing voting system. The occurrence of human errors is largely reduced due to the use of devices in automating the processes involved in the elections, for the proposed voting system.

6) Cost: The costs incurred is large, as a result of printing, transportation and safe-keeping of ballot papers in the existing system. But less expenses are made with the proposed voting system, due to the extreme reduction of paper usage, as devices which are used could be auctioned for sales after the election.

7) Collation of election results: Human errors are bound to occur while collating the results manually, and this tends to consume a lot of time in the existing system. The election results are collated by the proposed voting system immediately a vote is cast. The results collation is automated, thereby reducing the possibility of human errors and saving time.

V. CONCLUSIONS AND FUTURE SCOPE

Throughout the project, we have been able to develop a voting software which manages and maintains the voter's information and biometric data of the voters. Also the voting software eliminates the need for the printing and transportation of ballot papers to various designated polling stations, thereby removing any expenses that would have been caused as a result of this. Finally, this voting software does the task of automating the collation of results in real-time that is as the election is being conducted.

If given a second chance, the security features of this project would be greatly improved on, pertaining to both the hardware and the software, such as, the use of QR codes rather than barcodes.

If this project is to be worked on in the near future, a necessary factor we missed out in putting into consideration in this voting system, which should be included involves a voting system that can also be of service to people with disabilities, such as the blind. This is because they are also part of the society and voting is also their fundamental human right. Also, various technologies are available now and are still evolving. Alongside with this is vast growth in information and security risks or threats, to these technologies. This makes data security a high priority for this Biometric Voting System. There is the crucial need of continuous improvement of the security features of this voting system so as to make it robust against present day security challenges such as security breaches, attacks and denial of service. Therefore provisions and services should be made for such in the voting system by any other person who wish to take up this project in the near future.

VI. LIMITATIONS

- Significantly, inadequate supply of relevant materials such as books and current information relating to the project affected our findings and research, due to insufficient information regarding existing electronic voting systems.

- Additional external devices such as the barcode scanner and thumb-print/finger-print readers, were costly to obtain and complex in the installation and operation of the devices.

- During the testing process, the findings obtained does not provide information about how all voters will react to the Biometric Voting System and its processes. The results are only based on the participant's feedback, no observation actions were performed.

REFERENCES

[1] V. K. Yadav et al., "Kerberos Based Electronic Voting System", Special Issue of International Journal of Computer Applications (0975-8887), November 2012.

[2] A. Al-Ameen and S. Talab, "The Technical Feasibility and Security of E-Voting", Department of Information Technology, University of Neelain, Sudan, The International Arab Journal of Information Technology, Volume 10, No. 4, July 2013.

[3] F. A. Haziemeh *et al.*, "New Applied E-Voting System", Journal of Theoretical and Applied Information Technology, Vol. 25 No. 2, 31st March 2011.

[4] M. A. Musa and F. M. Aliyu, "Design of Electronic Voting Systems for Reducing Election Process", International Journal of Recent Technology and Engineering, Volume-2, Issue-1, March 2013.

[5] L. Constantin. Last modified (2014, May 12). Estonian Electronic Voting System vulnerable to attacks, researchers say [Online]. Available: http://www.pcworld.com/article/2154000/estonian-electronic-voting-system-vulnerable-to-attacks-researchers-say.html

[6] "Voluntary Voting System Guidelines", Election Assistance Commission, May 2005.

[7] D. Balzarotti et al., "Are Your Votes Really Counted? Testing the Security of Real-world Electronic Voting Systems", Computer Security Group, University of California, Santa Barbara, 2008.

[8] M. Stenbro, "A Survey of Modern Electronic Voting Technologies", Norwegian University of Science and Technology, Department of Telematics, June 2010.